T0225236

Cambridge Elements ≡

Elements in Geochemical Tracers in Earth System Science
edited by
Timothy Lyons
University of California
Alexandra Turchyn
University of Cambridge
Chris Reinhard
Georgia Institute of Technology

THE CHROMIUM ISOTOPE SYSTEM AS A TRACER OF OCEAN AND ATMOSPHERE REDOX

Kohen W. Bauer
University of Hong Kong

Noah J. Planavsky
Yale University, Connecticut

Christopher T. Reinhard
Georgia Institute of Technology

Devon B. Cole
Georgia Institute of Technology

CAMBRIDGE
UNIVERSITY PRESS

University Printing House, Cambridge CB2 8BS, United Kingdom

One Liberty Plaza, 20th Floor, New York, NY 10006, USA

477 Williamstown Road, Port Melbourne, VIC 3207, Australia

314–321, 3rd Floor, Plot 3, Splendor Forum, Jasola District Centre, New Delhi – 110025, India

79 Anson Road, #06–04/06, Singapore 079906

Cambridge University Press is part of the University of Cambridge.

It furthers the University's mission by disseminating knowledge in the pursuit of education, learning, and research at the highest international levels of excellence.

www.cambridge.org
Information on this title: www.cambridge.org/9781108792578
DOI: 10.1017/9781108870443

© Kohen W. Bauer, Noah J. Planavsky, Christopher T. Reinhard, and Devon B. Cole 2021

This publication is in copyright. Subject to statutory exception and to the provisions of relevant collective licensing agreements, no reproduction of any part may take place without the written permission of Cambridge University Press.

First published 2021

A catalogue record for this publication is available from the British Library.

ISBN 978-1-108-79257-8 Paperback
ISSN 2515-7027 (online)
ISSN 2515-6454 (print)

Cambridge University Press has no responsibility for the persistence or accuracy of URLs for external or third-party internet websites referred to in this publication and does not guarantee that any content on such websites is, or will remain, accurate or appropriate.

The Chromium Isotope System as a Tracer of Ocean and Atmosphere Redox

Elements in Geochemical Tracers in Earth System Science

DOI: 10.1017/9781108870443
First published online: January 2021

Kohen W. Bauer
University of Hong Kong

Noah J. Planavsky
Yale University, Connecticut

Christopher T. Reinhard
Georgia Institute of Technology

Devon B. Cole
Georgia Institute of Technology

Author for correspondence: Noah J. Planavsky, noah.planavsky@yale.edu

Abstract: The stable chromium (Cr) isotope system has emerged over the past decade as a new tool to track changes in the amount of oxygen in Earth's ocean-atmosphere system. Much of the initial foundation for using Cr isotopes (δ^{53}Cr) as a paleoredox proxy has required recent revision. However, the basic idea behind using Cr isotopes as redox tracers is straightforward – the largest isotope fractionations are redox-dependent and occur during partial reduction of Cr(VI). As such, Cr isotopic signatures can provide novel insights into Cr redox cycling in both marine and terrestrial settings. Critically, the Cr isotope system – unlike many other trace metal proxies – can respond to short-term redox perturbations (e.g., on timescales characteristic of Pleistocene glacial-interglacial cycles). The Cr isotope system can also be used to probe the Earth's long-term atmospheric oxygenation, pointing towards low but likely dynamic oxygen levels for the majority of Earth's history.

Keywords: chromium isotopes, redox, oxygenation, deoxygenation, metal mass balance

© Kohen W. Bauer, Noah J. Planavsky, Christopher T. Reinhard, and Devon B. Cole 2021

ISBNs: 9781108792578 (PB), 9781108870443 (OC)
ISSNs: 2515-7027 (online), 2515-6454 (print)

Contents

1 Introduction

There has been significant interest in recent decades in the use of transition metal isotope techniques to track the biogeochemical evolution of Earth's ocean-atmosphere system. In particular, transition metals, with measurable stable isotopic variability that can exist at a variety of redox states under Earth surface conditions (such as Cr, Fe, Mo, Tl, and U) have garnered much interest as potential tracers of the redox state of Earth's surface through geological time. This work has been driven by fundamental unresolved questions about past fluctuations in surface oxygen levels. For instance, surface warming over the next millennium is predicted to result in decreased levels of dissolved oxygen in seawater, potentially dramatically altering global biogeochemical cycles and reducing biological productivity and diversity in the world's oceans (Keeling et al., 2010). However, the extent of ocean deoxygenation during warming is still poorly constrained. Few studies have attempted to quantify the spatial extent of low oxygen and fully anoxic marine conditions on a global scale during past warming events. This has challenged efforts to gauge both the scope of redox-dependent feedbacks during various climatic perturbations, and the extent of future redox shifts. Equally, there is also intense debate about the magnitude and timing of larger redox shifts in deep time, including during all the major mass extinctions in Earth's history. Chromium isotopes are primed to become part of the standard toolkit – coupled with trace metal enrichments, nitrogen (N) isotopes, and other novel metal isotope systems – that we use to push forward our understanding of ocean deoxygenation.

There has also been persistent debate about Earth's long-term oxygenation over the past few decades (Lyons et al., 2014). This debate translates into significant uncertainty in the relative roles that environmental and biological factors have played in driving broad-scale evolutionary trends (Cole et al., 2020). For large, multicellular organisms like animals most of the debate has centered on whether oxygen levels were low enough to have prevented animals from establishing stable populations over million-year time scales (Butterfield, 2009; Erwin et al., 2011; Sperling et al., 2013; Towe, 1970). Quantitative constraints on oxygen levels during the majority of the Proterozoic – that is, the billion-year interval preceding the rise of animals, have been a key piece of information missing from this conversation (Kump, 2008; Lyons et al., 2014). Metal isotope systems – including Cr isotopes – can add to this debate by providing minimum or maximum constraints on oxygen concentrations at Earth's surface. Given that Cr redox cycling induces the largest fractionations in this isotope system, the extent of sedimentary Cr isotopic variability can be used to pinpoint when Cr oxidation "turned on" – locally or globally. The initiation of widespread Cr oxidation can, in principle, be quantitatively linked to a range of minimum atmospheric oxygen levels. The Cr isotope system is thus well-suited to answer questions about Earth's long-term evolution as well as recent oxygen dynamics in Earth's oceans.

Here we outline the basics of the Cr isotope system and the major remaining gaps in our knowledge of how this system works. We explore a few examples of how the Cr isotope system can provide unique insights into past fluctuations in Earth's oxygen levels. Rather than an exhaustive but cursory review of all the Cr isotope work from the past decade, we highlight a few case studies that illustrate how the Cr isotope system is well-suited to address questions regarding recent marine oxygen dynamics, as well as Earth's long-term oxygenation. We also highlight some pitfalls of previous Cr isotope work as a means of shaping future endeavors. The overall message is that although the Cr isotope system is much more complicated than originally envisioned, when applied thoughtfully, Cr isotopes remain an important part of the toolkit being used to track marine and atmospheric oxygen levels.

2 Basics of Cr Speciation and Isotope Fractionations

On the modern Earth's surface, the Cr cycle is largely governed by redox reactions whereby soluble Cr(VI) species are produced via oxidation of reduced Cr(III) species. Chromium is present almost exclusively in rock-forming minerals in igneous rocks containing reduced Cr(III) (Fandeur et al., 2009). Therefore, mobilization of Cr in soil environments necessitates the oxidation of Cr(III) to Cr(VI). However, because the kinetics of Cr(III) oxidation with O_2 are slow (Eary and Rai, 1987; Johnson and Xyla, 1991), Mn(III, IV) (oxyhydr-) oxides are typically considered to be the only environmentally relevant oxidant for Cr(III) at the Earth's surface (Bartlett and James, 1979; Eary and Rai, 1987; Fendorf, 1995). The role of superoxide, which has recently been highlighted as an important oxidant, for other metals has not been thoroughly investigated. Oxidation proceeds through dissolved Cr(III) reaction with solid phase Mn oxides and forms tetrahedrally coordinated oxyanions (e.g., CrO_4^{2-}, $HCrO_4^-$, $Cr_2O_7^{2-}$), which are highly soluble and readily transported in oxidizing aqueous fluids. Riverine chromate (CrO_4^{2-}) is thus considered to be the main source of Cr to the modern oceans (Bartlett and James, 1979; Fendorf, 1995; Konhauser et al., 2011; Oze et al., 2007). However, many rivers also contain a significant Cr(III) load (e.g., D'Arcy et al., 2016; Wu et al., 2017), which is likely bound by organic ligands.

In modern oxygenated oceans, Cr is thermodynamically stable and present predominantly as Cr(VI), although significant portions of Cr(III) exist in regions; for example, in some North Pacific water masses (Janssen et al., 2020; Wang et al., 2019). The generally conservative behavior of Cr in oxic waters is in strong contrast to that observed in anoxic systems. In anoxic systems Cr(VI) is quickly reduced by Fe(II), sulfide, solid-phase reduced Fe(II) and S phases and even some organic compounds when these are present at high concentrations (Eary and Rai, 1989; Fendorf, 1995; Patterson et al., 1997; Richard and Bourg, 1991). On reduction at circumneutral or alkaline pH, the majority of resulting Cr(III) will hydrolyze to form $Cr(OH)_3$, which is sparingly soluble and readily removed from solution. Therefore, within any anoxic aquatic system, the Cr reservoir will be largely reduced to Cr(III) and subsequently preferentially partitioned into solid (particulate) phases.

Chromium isotope compositions are reported in delta notation ($\delta^{53/52}Cr$), relative to the international standard NIST SRM-979 ($\delta^{53}Cr = 1000 \times [(^{53}Cr/^{52}Cr)_{sample}/(^{53}Cr/^{52}Cr)_{SRM-979} - 1]$). Early work on the Cr isotope system in Cr(VI)-contaminated groundwaters established the view that surface water $\delta^{53}Cr$ values were controlled by redox transformations (Johnson and Bullen, 2004). Both theoretical and experimental studies indicate that Cr will undergo limited fractionation during most redox-independent transformations, for example, adsorption processes (Ellis et al., 2004; Johnson and Bullen, 2004; Schauble et al., 2004). The typically limited extent of redox-independent Cr isotope fractionations is largely linked to bonding preferences. Chromium(III) has a very strong preference for octahedral coordination, while Cr(VI) strongly favors a tetrahedral coordination. This is in contrast to other heavy metal isotope systems where, in a given redox state, both tetrahedral and octahedral coordination environments are common (e.g., Fe, Cu). As a result, non-redox-linked coordination changes are less likely to drive isotope fractionations of Cr than in "typical" heavy metal isotope systems (Schauble et al., 2004). Chromium(III) complexation with ligands is the most notable exception to this rule (e.g., Babechuk et al. (2018); Saad et al. (2017)). For instance, chromium(III)-organic ligand complexation can cause large (> 1‰) fractionations, but the exact mechanism driving this fractionation is currently unresolved. Saad et al. (2017) suggested that the isotope fractionation occurs during a back reaction, but additional experimental work and ab initio calculations are needed for a more detailed mechanistic understanding of most redox-independent Cr isotope fractionations (see Babechuk et al., 2018).

In marked contrast to most non-redox processes, the oxidation and reduction of Cr species induce large isotope fractionations. Because there is a narrow range of $\delta^{53}Cr$ values observed in igneous systems, with an average value of –0.124 ±0.101‰ (2SD) (Schoenberg et al., 2008), Cr isotope fractionations are often compared to this crustal range, which is canonically referred to as the igneous silicate Earth (ISE) composition. At equilibrium, the $Cr(VI)O_4^{2-}$ anion is predicted to be enriched in heavy Cr (^{53}Cr) by over 6‰ relative to the coexisting Cr(III) reservoir (Schauble et al., 2004). However, in natural Earth surface systems it is unlikely that the full equilibrium isotope effect will be expressed. Fractionations resulting from oxidation in natural systems observed thus far are less than 1‰. In contrast, fractionations observed during reduction from Cr(VI) to Cr(III) range between 3‰ and 5.5‰ (Ellis et al., 2002, 2004; Johnson and Bullen, 2004; Schauble et al., 2004; Zink et al., 2010). However, if reduction is quantitative, as would be expected in strongly reducing environments, large fractionations will not be expressed at the system scale (Reinhard et al., 2013).

3 A Global Cr Isotope Mass Balance?

3.1 Marine Cr Input Fluxes

The discharge-weighted riverine input of dissolved Cr to the ocean was estimated by (Reinhard et al., 2013) to be ~6 x 10^8 mol y^{-1}. This estimate is significantly below that of McClain and Maher (2016), who estimated a Cr input flux roughly a factor of two higher (~1.7 x 10^9 mol y^{-1}). However, recent work on catchments with minimal anthropogenic influence revealed anomalously low Cr concentrations (~4 x 10^4 mol y^{-1}) relative to those from similar climate zones in the McClain and Maher (2016) study (roughly an order of magnitude lower Cr concentrations (Wu et al., 2017)). Significant anthropogenic riverine Cr contamination seems likely, with the result that accurately reconstructing the pre-anthropogenic riverine Cr flux is challenging. The potential for anthropogenic Cr contamination in rivers also affects our ability to estimate average pre-anthropogenic riverine $\delta^{53}Cr$ values. Nonetheless, it is interesting that riverine $\delta^{53}Cr$ values are highly variable, but mostly enriched towards heavy compositions relative to crustal values. Most recently, the average riverine $\delta^{53}Cr$ value was estimated to be roughly +0.47 ± 0.39‰ (Toma et al., 2019).

Hydrothermal systems do not appear to be a large Cr source, in contrast to many transition metals. High-temperature hydrothermal fluids may be depleted in Cr relative to seawater due to early mixing of hydrothermal fluids and seawater and the rapid formation of Fe (oxyhydr-)oxides, which remove Cr through co-precipitation and scavenging (German et al., 1991). Furthermore, fluid concentration anomalies combined with estimates of global heat flux associated with axial hydrothermal activity (Elderfield and Schultz, 1996) indicate that these systems amount to a small net sink of Cr from seawater (Reinhard et al., 2013). This is corroborated by low Cr inventories in sediments underlying hydrothermal systems (Bauer et al., 2019). The effect of diffuse-flow hydrothermal systems is poorly constrained, although there is some evidence that low-temperature hydrothermal fluids are only mildly enriched in Cr relative to seawater (Sander and Koschinsky, 2000). However, if these concentration anomalies are extrapolated globally by assuming the entire riverine Mg^{2+} flux is removed in diffuse-flow systems, the estimated Cr flux would be ~3–4 x 10^6 mol y^{-1}, still less than ~1% of the dissolved riverine flux (Reinhard et al., 2013). We also note that measurements to date indicate that the δ^{53}Cr of hydrothermal fluids is close to the ISE value (Bonnand et al., 2013).

3.2 Marine Cr Burial Fluxes and Isotope Mass Balance?

There are four primary marine Cr removal fluxes (Figure 1): (1) burial in sediments deposited within anoxic water columns; (2) burial in reducing continental margin sediments; (3) burial in carbonate depositional environments; and (4) burial in oxic marine sediments. Removal of Cr from seawater in any of

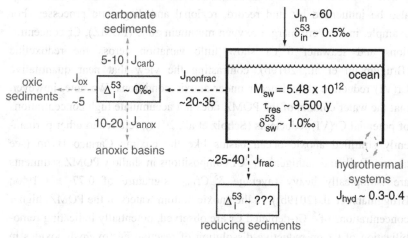

Figure 1 Schematic of the provisional global Cr mass balance discussed in the text and revised from (Reinhard et al., 2013). The magnitude and isotopic composition of the different Cr fluxes have been updated based on Cr studies conducted in the modern oceans (Gueguen et al., 2016; Sander and Koschinsky, 2000; Sander et al., 2003; Scheiderich et al., 2015; Toma et al., 2019). Seawater Cr reservoir mass (M_{sw}), oceanic residence time (τ_{res}), and isotope composition (δ^{53}_{sw}) are shown, as are the isotopic offsets from seawater associated with burial in each sink (Δ^{53}_i terms). Numbers associated with each flux (J_i) are in units of 10^7 mol y^{-1}. Note that the separation of sink terms into "fractionating" (J_{frac}) and "non-fractionating" ($J_{nonfrac}$) is provisional

these environments could, in principle, be accompanied by an isotopic fraction-ation – causing seawater δ^{53}Cr values to deviate from the mean time-integrated input value.

3.2.1 Anoxic Sediments

Chromium(VI) may be efficiently scavenged from anoxic water columns, and thus Cr removal in anoxic waters likely represents a significant removal flux despite the small spatial extent of these environments. Work on sediments deposited in the Cariaco Basin, a permanently anoxic marine basin off the coast of Venezuela (Gueguen et al., 2016; Reinhard et al., 2013), suggests that authigenic Cr in euxinic (anoxic, H_2S-rich) sediments is isotopically similar to contemporaneous seawater δ^{53}Cr values (Figure 2). The δ^{53}Cr values are between 0.6‰–0.7‰, which is within error of adjacent Atlantic seawater values (Bonnand et al., 2013). The authigenic Cr isotope composition of Cariaco Basin sediments determined using a detrital element correction (using a Cr/Ti ratio) and a set of weak acid leaches are comparable (Figure 2).

The δ^{53}Cr composition of Cr buried in some anoxic marine sediments may also be influenced by, and record, regional and diagenetic processes. For example, in the Peru margin oxygen minimum zone (POMZ), Cr concentra-tions and seawater δ^{53}Cr show little variation across the redoxcline (Bruggmann et al., 2019b), contrasting the view that near quantitative Cr(VI) reduction should occur under anoxic conditions. We note, however, that the water column of the POMZ does not accumulate high concentrations of potential Cr(VI) reductants (Scholz et al., 2014), relative to other perman-ently stratified anoxic ocean basins like the euxinic Cariaco Basin (see above). Bulk and authigenic δ^{53}Cr compositions in shallow POMZ sediments are isotopically heavy (average δ^{53}Cr$_{Bulk}$ signature of 0.77 ± 0.19‰; Bruggmann et al. (2019b)). In the anoxic bottom waters of the POMZ, higher concentrations of ^{53}Cr-depleted Cr are observed, potentially indicating remo-bilization of Cr on reductive dissolution of reactive Fe (oxyhydr-)oxides in surface sediments. Bruggmann et al. (2019b) also observe covariation between δ^{53}Cr and organic matter contents in the POMZ sediments, which may be indicative of a small yet important authigenic Cr pool delivered to the sediment surface associated with plankton biomass. Furthermore, this organic- and Cr-rich end member has a heavy Cr isotopic composition (1.10 ± 0.08‰). Chromium isotope data from the POMZ thus reveals that even under prevailingly anoxic water column conditions, regional factors such as organic matter loading and the delivery and speciation of Fe may give rise to δ^{53}Cr variations recorded in the sedimentary record.

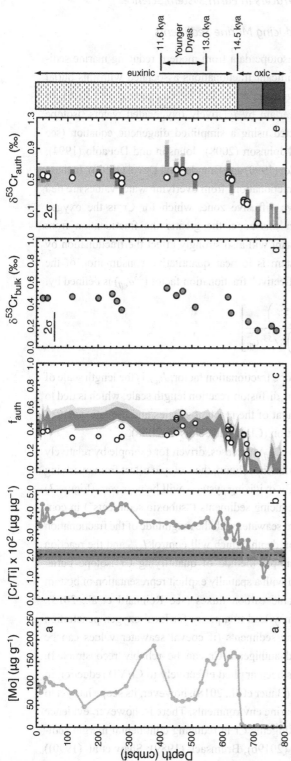

Figure 2 Geochemical data for sediments deposited at ODP Site 1002 since the Last Glacial Maximum (reproduced from Reinhard et al., 2014). Core stratigraphy at right and data for Mo in (**a**) are from Lyons et al. (2003). Gray field in (**b**) denotes the range of Cr/Ti values for upper continental crust (UCC), while the black dotted line and red field denote the mean and 95% confidence interval for sediments deposited under oxic conditions. Values for the authigenic Cr fraction (f_{auth}; **c**) are calculated based on measured [Cr/Ti] values and assuming the resampled mean and 95% confidence interval of oxic sediments as the detrital background (red field) or the range of estimates for UCC (gray field: Rudnick (2003)). In (**c**), are values calculated based on leachate [Cr] data (open circles). The Cr isotope composition of bulk sediments is shown in (**d**). In (**e**), the isotopic composition of the reconstructed authigenic component ($\delta^{53}Cr_{auth}$) is shown, according to calculations using the UCC composite range (gray bars), the mean and 95% confidence interval for oxic sediments (red bars), and the sediment leaches (open circles). The blue shaded region denotes the range for modern deep Atlantic seawater (Bonnand et al., 2013). Error bars in (**d**, **e**) show external reproducibility (2σ) of ±0.1‰

3.2.2 Reducing Marine Sediments

There is limited published Cr isotope data from modern reducing marine sediments, however, there are significant fractionations associated with this burial term (Bauer et al., 2018; Bruggmann et al., 2019b; Gueguen et al., 2016). In reducing marine sediments overlain by relatively oxygenated waters, isotope fractionations are often predicted using a simplified diagenetic equation (see Bauer et al. (2018); Clark and Johnson (2008); Johnson and DePaolo (1994); following the approach of Bender (1990) and Brandes and Devol (1997)). In these sedimentary systems, isotope fractionations from overlying water values are tied to the scale of the nonreactive diffusive zone, which for Cr is the oxygen penetration depth. The thinner this zone, the greater the fractionation from overlying water. The rate constant will also strongly affect the fractionation by controlling how close the system is to near quantitative consumption of the porewater Cr reservoir. The "effective" fractionation factor ($^{53}\alpha_{eff}$) is defined by:

$$^{53}\alpha_{eff} = \sqrt{^{53}\alpha_{int}} \left[\frac{1 + \left(\frac{L_{diff}}{\lambda}\right)}{1 + \left(\frac{L_{diff}}{\lambda}\right)\sqrt{^{53}\alpha_{int}}} \right]$$

where $^{53}\alpha_{int}$ is the intrinsic isotope fractionation factor; L_{diff} is the length scale of the nonreactive zone and λ is the diffusion-reaction length scale, which is tied to the sediment diffusion coefficient of the reactant species and the rate constant for a first-order reduction reaction (Clark and Johnson, 2008).

As the diffusive length scale (L_{diff}) increases, driven for example by relatively deep O_2 penetration into the sediment column, the isotopic offset from overlying water decreases, and for strongly oxidizing systems will be near zero (Figure 3) (Clark and Johnson, 2008). Reducing sediments ("suboxic sediments") in contrast should be fractionated from seawater with the magnitude of the fractionation scaling with oxygen penetration depth, which will control L_{diff} and the reaction rate term. In principle, this simple method of quantifying Cr isotope burial fractionations could be coupled with a spatially explicit representation of bottom water oxygen levels and organic carbon fluxes (see Reinhard et al., 2013). Additionally, this framework may lead to a proxy for local oxygen penetration depths in past reducing marine sediments (if coeval seawater values can be independently constrained and authigenic Cr can be reliably reconstructed).

The idealized 1D model has been applied effectively to Cr(VI) reduction in lacustrine-reducing sediments (Bauer et al., 2018), however, its scope has yet to be fully explored in modern marine environments. There is, however, evidence for elevated burial rates of authigenic Cr in reducing continental margin sediments (e.g., Bruggmann et al. (2019b); Brumsack (1989); Shaw et al. (1990))

Table 1 Trace element ratio
ranges indicative of pyrite formed
in sedimentary settings (Gregory
et al., 2015)

Ratio
$0.01 < Co/Ni < 2$
$0.01 < Zn/Ni < 10$
$0.01 < Cu/Ni < 2$
$0.1 < As/Ni < 10$
$1 < Te/Au < 1000$
$As/Au > 200$
$Ag/Au > 2$
$Sb/Au > 100$
$Bi/Au > 1$

Gregory et al., 2017). This technique has since been refined by coupling a database of known pyrite (five types of hydrothermal pyrite and sedimentary pyrite) trace element content with machine-learning algorithms. By doing this, sedimentary pyrite can be correctly identified 95 percent of the time per analysis and 100 percent of the time per sedimentary formation (Gregory et al., 2019a).

3 Materials and Methods

3.1 Pyrite Textures and Sample Preparation

Pyrite commonly forms in the pore waters of organic rich sediments or in sulfidic water. As both of these settings tend to be organic-rich, black shales are the main rock types in which the pyrite trace element proxy can be utilized. As with most geochemical proxies, the pyrite proxy gives only local conditions; to make sweeping generalizations about global ocean chemistry, samples from multiple locations representative of the same time interval should be obtained (Large et al., 2015).

The in-situ nature of the pyrite proxy technique simplifies the sample preparation procedures compared to many other proxies. No grinding and dissolution is required. The samples only need to be cut, mounted in epoxy, and polished with a final grit of 1 μm diamond paste. Extra care should be taken to avoid over-polishing, which can result in an uneven surface due to soft material being preferentially polished away. This is because uneven surfaces can result in inefficient transport of the ablated material into the ICPMS, which may adversely affect the LA-ICPMS analysis. Once polished mounts have been

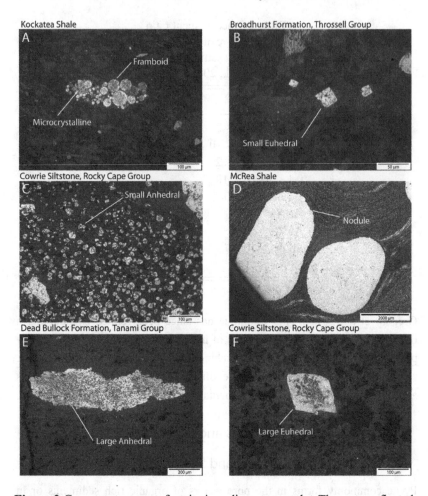

Figure 2 Common textures of pyrite in sedimentary rocks. These are reflected light images and samples have been etched with nitric acid. A) framboidal pyrite, B) small (< 15 μm) euhedral pyrite, C) small (< 15 μm) anhedral pyrite, D) nodular pyrite, E) large (> 15 μm) anhedral pyrite, F) large (> 15 μm) euhedral pyrite. Figure from Gregory et al., (2015a).

made, detailed petrographic observation should be undertaken to identify pyrites that are likely to have formed during the stage in the history of the sediments that is of interest. Nitric acid etching (Gregory et al., 2015a; Fig. 2, amongst many others) or NaOCl (Sykora et al., 2018) can be very useful in revealing the textures of the initial pyrite generations, as well as the relative timing of the formation of the different textured pyrite. Small (< 10 μm) pyrite framboids (Fig. 2A) are the most likely to have formed in the water column (Wilkin et al., 1996); thus, they are also the most likely to accurately reflect the trace element content of the ocean water they formed in and are the preferred textures to analyze. However, those framboids are difficult to analyze due to their small size, so clusters can be analyzed wherever available. Sometimes early, fine-grained pyrite framboids have later pyrite overgrowths (Wacey, et al., 2015; Gregory et al., 2019b); if such an early texture can be identified from later pyrite, they are also good targets for analysis. Late pyrite tends to be coarser grained and/or more euhedral, and are best avoided for past ocean chemistry studies.

If the goal is rather to track the changes through sediment diagenesis, analysis across later, rimming pyrite can give some useful information. This application again begins with detailed petrography including etching. Then either LA-ICPMS maps (Large et al., 2009; Genna and Gaboury, 2015, Gregory et al., 2015a; Steadman et al., 2015; Cooke et al., 2016) or a line of spot analyses (Gregory et al., 2019b) are conducted across the different identified textures to give an understanding of how pyrite chemistry (and by inference, pore fluid chemistry) varied during the formation of the pyrite. By coupling these analyses with in-situ S-isotope (SIMS or SHRIMP) analysis, the degree of openness of the system can be determined and an understanding of which diagenetic processes were occurring can be achieved.

3.2 Analytical Methods: LA-ICPMS

LA-ICPMS is now a common technique in many aspects of earth sciences, including dating of different mineral phases, understanding changes in chemistry of individual igneous and metamorphic minerals, determining fluid inclusion chemistry, understanding the chemistry of ore minerals, and now pore/ocean water chemistry via pyrite chemistry. Here, a summary of the method is presented and some potential issues and best practices for reporting data are highlighted. Interested readers are referred to Large et al. (2007), (2009), and Gregory et al. (2017), and references therein for details on the methodology of LA-ICPMS analysis of sedimentary pyrite.

The LA-ICPMS technique, essentially, utilizes laser energy to vaporize a small amount of the mineral of interest (in this case pyrite) in a He atmosphere. The resultant aerosol is then carried (an additional carrier gas, usually Ar, is usually added to help aerosol transport) to the mass spectrometer where it is ionized in a plasma and analyzed by the mass spectrometer. In addition to the target pyrite grains, approximately five spot analyses can also be analyzed on the black shale matrix, taking care to avoid small pyrite grains. This can later be used to correct for the matrix material that is often ablated with the target pyrite, especially with fine-grained pyrite. At the beginning and end of each set of analyses, a standard or multiple standards with known major and trace element composition is analyzed. The counts of the standard are compared to counts for an internal standard to convert counts of measured elements in the unknown pyrite to ppm (Fe is the most appropriate for pyrite; Longerich et al., 1996). Secondary standards, ideally pyrite or similar sulfide material, should also be interspersed with the unknown samples to check the accuracy of the analytical run. In addition to the elements of interest, major elements are also monitored to determine the quantity of matrix inclusions and appropriate standards to properly quantify them are utilized.

3.2.1 Data Processing

Once initial data reduction has been completed, the total abundance should be checked to ensure that the sum of all the elements is near 100 percent, using masses of the elements related to their most likely species. That is, you need to correct for elements that likely exist bound to oxygen as oxygen will not be quantified. If necessary, a correction factor should be applied to the resultant data. Once total abundances are calculated, the data are scrutinized for the amount of non-pyrite matrix minerals. It is recommended that if an analysis captured more than 20 percent matrix it is excluded; and that analyses that have less than 20 percent matrix are adjusted by using a correction factor that accounts for the trace elements held within the matrix inclusions. University of Tasmania researchers Leonid Danyushevsky and Sasha Stepanov specialize in LA-ICPMS data reduction algorithms and they suggest segmenting the area of integration during data processing into five groups. The chalcophile and siderophile elements are then plotted against S and linear regression equations are calculated. The final concentrations are then calculated with the S content that brings the total siderophile and chalcophile concentration up to 100 percent (see Gregory et al., 2017; Stepanov et al., 2020, for details).

In addition to understanding the changing trace element concentrations captured by sedimentary pyrite, the data obtained through LA-ICMPS can also shed light on how trace elements are held in pyrite (i.e., as micro-inclusions or bound within the pyrite structure). This is done by close inspection of time resolved laser ablation output graphs (Fig. 3). When there are irregular sharp peaks or drops in the total counts, it is likely that the laser beam is ablating through a single or multiple inclusions. An example of this is the two Au peaks in Fig. 3D and Pb and Bi peaks in Fig. 3E and 3 F respectively. All of these peaks are likely due to the laser beam ablating through micro-inclusions of minerals that contain the respective elevated elements in them. In contrast, Ni, Co, and As tend to parallel the counts of Fe in each of the examples in Fig. 3, indicating

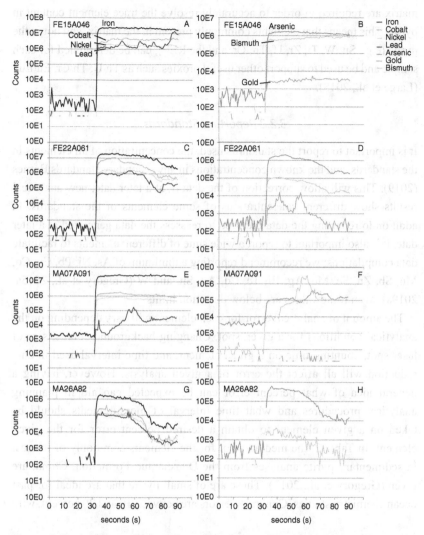

Figure 3 Typical time resolved laser ablation output graphs from sedimentary pyrite from 4 different samples. Most trace elements are relatively even counts throughout ablation, indicating consistent trace element content. A few analyses have peaks in Pb and Bi (e and f) indicating an inclusion of a mineral containing both these elements and two peaks in Au are evident in d, indicating inclusions of Au. Figure from Gregory et al., (2015a).

that these minerals are held within the pyrite structure. However, LA-ICPMS is not considered sensitive enough to determine whether a given element is held with the structure of pyrite or within evenly distributed nano-inclusions (Gregory et al., 2015a). As outlined, LA-ICPMS analyses of the sedimentary matrix are required in order to accurately resolve the trace element content in pyrite. This means that elements commonly present in the matrix, and not in the pyrite (e.g., Sn, W, Ti, Zr, U, Th, Cr, and the REE), can be determined for each sample and be used to support other useful proxies such as Th/U, Th/Cr, Eu/Eu* (Large et al., 2018).

3.2.2 Reporting Standards

It is important to report the standards used, the concentration of all elements in the standards, and the known concentrations in any secondary standards Barnes (2019). This will allow correction of the data if, at a later date, new analytical results show different concentrations of some elements of the standards. In addition to reporting the data required to reassess the data generated at a later date, it is also important to report a wide suite of different elements to facilitate data compilations: we recommend reporting a minimum of As, Ni, Pb, Cu, Co, Mn, Sb, Zn, Se, Mo, Ag, Bi, Te, Cd, Au, Sn, and W (Gregory et al., 2015a; 2019a), even if the values are below detection limits.

The amount of analytical error for each element will vary depending on the analytical conditions for a given spot. Changing background counts, size of laser spot, counting time on a given element, and time interval used in data reduction will all affect the error of a given analysis. However, having a general idea of what percent error can be expected can aid in planning analytical procedures and what time interval of measurements should be taken on a given element to obtain a desired percent error for the given element. In Table 2, the median error for the elements commonly of interest in sedimentary pyrite analyses from the Doushantuo Formation, China, are given (Gregory et al., 2017). These are of small pyrite that are ideal for past ocean chemistry studies. Generally, errors are between 10 and 20 percent, with

Table 2 Median % errors for elements commonly held in pyrite from samples analyzed in Gregory et al., 2017

Element	Mn	Co	Ni	Cu	Zn	As	Se	Mo	Ag	Sb	Te	Au	Tl	Pb	Bi
Median % error	27	8.9	6.5	12	60	6.6	7.2	18	12	9.5	15	19	17	9.1	14
Associated concentration* (ppm)	118	50	300	522	39	250	61	9.3	3.9	40	2.8	0.08	2.5	116	2.3

* By associated concentration we mean the concentration of the given element for the sample with the median error estimate. This is done because error estimates are related to the total counts for the element of interest.

the higher errors usually belonging to elements with lower abundance. For example, Au has a median % error of 19 percent but the sample with this error had a measured value of 0.08 ppm Au. Conversely, Ni has a low median % error (6.5) but with a relatively high concentration (300 ppm). Some elements do deviate from this trend, such as Zn (median % error 60) and Mn (median % 27). This may be due to these elements often being contained within micro-inclusions (Gregory et al., 2015a) in pyrite resulting in variable concentrations being integrated. Thus particular care must be taken when interpreting results from these elements.

3.2.3 Potential Issues with LA-ICPMS Analyses

One of the benefits of LA-ICPMS analyses is that it does not require dissolution of the sample and thus difficulties in dissolving geologic samples and the dilution that accompanies any dissolution can be avoided. However, this also means that no separation chemistry can be conducted prior to analysis in the ICPMS, thus possibility of mass peak overlaps can be high and elements that could potentially cause peak overlaps on elements of interest should be monitored. For example, $^{103}Rh^+$ has a mass charge ratio of +103, but so do $^{206}Pb^{2+}$, $^{87}Rb^{16}O^+$, $^{87}Sr^{16}O^+$, $^{87}Rb^{16}O^+H^+$, and $^{63}Cu^{40}Ar^+$ (Danyushevsky, 2019) – so care must be taken when interpreting the results of LA-ICPMS analyses. Because every sample will have different abundances of the different elements, it is difficult to predict potential interferences and analysts are encouraged to investigate pertinent ICP-MS literature, such as May and Wiedmeyer (1998), and also critically examine the potential interferences in their unique samples prior to analysis. For paleo-ocean chemistry studies, As, Mo, Co, Ni, and Se can be among the most important and each has its own potential interferences. Selenium can be particularly problematic because ^{74}Se, ^{76}Se, ^{78}Se, and ^{80}Se, which include the most abundant Se isotopes (^{78}Se abundance of 23.53 percent and ^{80}Se abundance of 49.82 percent) and all have interferences with different polyatomic Ar species. For example, ^{74}Se is interfered by $^{36}Ar^{38}Ar^+$, ^{76}Se is interfered by $^{40}Ar^{36}Ar^+$ and $^{38}Ar^{38}Ar^+$, ^{78}Se is interfered by $^{40}Ar^{38}Ar^+$, and ^{80}Se is interfered by $^{40}Ar^{40}Ar^+$ (Tan and Horlick, 1986; Longbottom et al., 1994; May and Weider, 1998). These can be particularly problematic because Ar is a carrier gas in most LA-ICPMS experimental apparatus and thus production of these species can be difficult to avoid. Nickel also has potential interferences on Ar containing polyatomics such as $^{40}Ar^{18}O^+$ and $^{40}Ar^{17}O^1 H^+$ on ^{58}Ni. For Ni there may also be interferences from polyatomic molecules that may come from matrix material in the rock samples such as $^{23}Na^{35}Cl^+$, $^{40}Ca^{18}O^+$, $^{40}Ca^{18}O^1 H^+$, and $^{42}Ca^{16}O^+$ for ^{58}Ni and $^{23}Na^{37}Cl^+$, $^{44}Ca^{16}O^+$, and $^{43}Ca^{16}O^1 H^+$ for ^{60}Ni

(McLaren et al., 1985; Tan and Horlick, 1986; Plantz et al., 1989; Evans and Giglio, 1993; Reed et al., 1994; May and Weidmeyer, 1998). These could be particularly important interferences in modern ocean sediments for NaCl or carbonate rich sediments for the CaO interferences. Molybdenum has a number potential interference for each of its isotopes that could be derived by phyllosilicate minerals in the matrix material. These include $^{39}K_2^{16}O^+$, for ^{94}Mo; $^{40}Ar^{39}K^{16}O^+$ for ^{95}Mo; $^{39}K^{41}K^{16}O^+$ for ^{96}Mo; $^{40}Ar^{41}K^{16}O^+$ and $^{40}Ca_2^{16}O^1\ H^+$ for ^{97}Mo; and $^{41}K_2O^+$ (Beary and Paulson, 1993; Vandecasteele et al., 1993, May and Weidmeyer, 1998). One possible strategy for determining whether there is a significant effect on the abundance of one element is by monitoring its other isotopes because, while they may all be affected as shown, the isotopes should not be equally affected so if each isotope gives the same value for its elements abundance it can be accepted as being reasonably accurate. This will not work for elements such as Co and As as they only have one isotope. Like Ni, Co and As can have interferences from carbonate rich matrixes because there are interferences from $^{43}Ca^{18}O^+$ and $^{42}Ca^{18}O^1\ H^+$ on ^{59}Co and $^{43}Ca^{16}O_2^+$ on ^{75}As (Tan and Horlick, 1986; Evans and Giglio, 1993; Campbell et al., 1994; Longbottom et al., 1994). The risk of incorrectly identified elements can be limited by monitoring the potentially interfering elements and checking whether increases in one are coupled with increases in the other, which could show an interference rather than a real elemental enrichment.

Due to their fine-grained nature, black shales can have abundant volatile molecules adsorbed to them and the volatiles can desorb during LA-ICPMS analysis. This can lead to high background counts and some unanticipated peak overlaps. To limit this effect, samples and standards are recommended to be kept under vacuum prior to analysis (Danyushevsky, 2019). If time between sample preparation and analysis is limited, then time in a low temperature vacuum oven can speed up this process.

3.2.4 Summarizing Data

Pyrite trace element content has variable distributions. It is almost never normally distributed (Gregory et al., 2019a), and, as such, should not be summarized using arithmetic means and standard deviations. More commonly, trace element content approximates a log normal distribution. If this is the case, then the geometric mean and multiplicative standard deviation can be used effectively. However, it is important to understand the statistical distribution of a dataset. Check whether there is indeed a log normal distribution, as this is not always the case, especially with low abundance elements, such as gold. If the data is not log-normally distributed, then the median value and the median absolute deviations can be used instead (Gregory et al., 2019a). We stress that the use of the correct statistical quantities is very important when reporting LA-ICPMS analyses of pyrite because it is not uncommon to have a multiple order of magnitude spread in the data. This means that single far outliers can significantly skew the mean and the calculated arithmetic standard deviation will represent far less than 68 percent of the data and/or the interval defined by the standard deviation will go into the negative, a physical impossibility.

4 Case Study

4.1 Comparison between Pyrite Trace Element Chemistry and Traditional Paleo-Ocean Proxies

To test the pyrite trace element proxy at the section scale, Gregory, et al. (2017) selected the Wuhe section of the Doushantuo Formation, China. This section was chosen because it was deposited under persistently euxinic conditions as determined by a combination of Fe-speciation analyses (Sahoo et al., 2016) and pyrite framboid size analysis (Wang et al., 2012). The well-established depositional setting eliminates the potential complexity of changing water column redox conditions. Additionally, the section had already been analyzed by Sahoo et al. (2016) for whole rock redox sensitive trace element abundance (i.e., Mo, V, U, and Re) and these data had been interpreted to show three distinct short-lived oxygenation events prior to the NOE (see Tribovillard et al., 2006, for a description of these whole rock proxies). Sahoo et al., (2016) also presented $\delta^{34}S$ values for the section, which informs on how the sulfate budget changed during deposition of the Wuhe section.

Gregory et al. (2017) found that most of the chalcophile elements are elevated in pyrite at the same horizon as the redox sensitive elements are elevated in whole rock. Importantly, the pyrite proxy also appeared to be more sensitive than the whole rock proxy, with Mo enrichment factors of 6.3 to 23.9 in intervals interpreted to have experienced oxygenation events (Fig. 4). This provides support for the conclusions of Sahoo et al., (2016) and also provides support for the use of the pyrite trace element proxy for sedimentary sections deposited under euxinic conditions. A larger number of chalcophile elements than expected were found to correctly predict periods of increased oxygenation. This is interpreted to be because of the chalcophile elements' affinity for bonding with sulfur. During times of large amounts of euxinia in the basin, pyrite is formed over a wide area resulting in widespread drawdown of chalcophile elements and a general depletion in these elements in the basin as a whole (Gregory et al., 2017). The relatively high $\delta^{34}S$ supports this interpretation as at times of widespread pyrite formation sulfur becomes depleted and residual sulfate becomes more enriched in the heavy

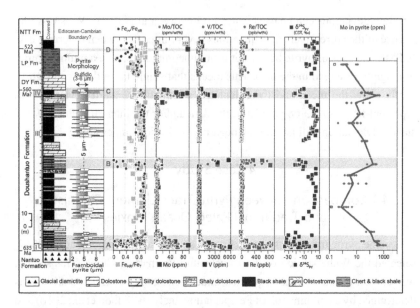

Figure 4 Comparison of whole rock geochemistry (Sahoo et al., 2016) and pyrite trace element content at the section scale. Note the increase of Mo and most of the chalcophile elements during the same intervals that there are increases of redox sensitive trace elements. This demonstrates how Mo and chalcophile trace element composition of pyrite can be used in a similar fashion to bulk analyses of redox sensitive trace elements. Figure from Gregory et al., (2017).

[34]S because as the lighter [32]S is preferentially used up. Then at times when oxygen increases the area of the seafloor under an euxinic water column decreases which limits production of pyrite and slows the drawdown of both [32]S and chalcophile trace elements (Gregory et al., 2017, Sahoo et al., 2016). Thus where the water column is still euxinic, chalcophile trace elements in pyrite increase and δ^{34}S decreases. Therefore, not only does the pyrite proxy reflect traditional proxies, it allows far more trace elements to be utilized in paleo-ocean chemistry reconstructions (Gregory et al., 2017).

5 Future Prospects

While promising, there are still many unresolved issues with the pyrite trace element proxy. Most of the studies thus far have focused on producing compilations of pyrite data through geologic time (Large et al., 2014; Gregory et al., 2015a; Large et al., 2015). While these compilations tend to match well with existing whole rock chemistry compilations (Scott et al., 2008; Konhauser et al.,

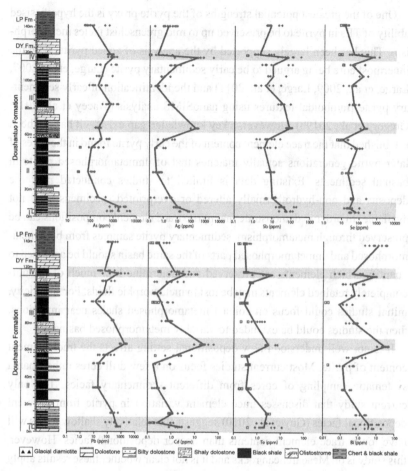

Figure 4 (cont.)

2009; Partin et al., 2013), there are still many questions that are unresolved. First, pyrite can form in many environments, within sediments under oxic or anoxic water columns or in the water column in euxinic settings. To what degree the trace element content of pyrite varies in each of these environments and how pyrite chemistry should be compared when sourced from different environments remains untested. Detailed analyses of pyrite forming in modern settings still need to be conducted to fully understand the potential variability of pyrite formed under these different conditions and to produce a work flow to refine the proxy. Furthermore, these studies should be conducted through ocean drilling program cores to ascertain how the trace element content of pyrite changes as different sulfate reduction mechanisms become more prevalent (i.e., thermochemical sulfate reduction).

One of the greatest potential strengths of the pyrite proxy is the hypothesised ability of TEs in pyrite to be preserved up to mid greenschist facies metamorphism. This has been largely supported by the analysis of zoned pyrite, with the innermost core being argued to be early sedimentary pyrite (Large, et al., 2007, Large, et al., 2009, Large, et al., 2011) and the identification of early sedimentary pyrite framboidal textures using nanoSIMS analysis (Wacey et al., 2015; Gregory et al., 2019b). However, a key knowledge gap exists – it has not been established that the trace element content of the early pyrite found in the cores of later pyrite generations actually matches that of unmetamorphosed pyrite in basinal sediments. Existing data is limited to studies conducted near ore deposits and non-hydrothermally altered or overprinted specimens were not available. In order to demonstrate that pyrite trace element compositions can be preserved through metamorphism, sedimentary pyrite samples from both metamorphosed and unmetamorphosed parts of the same basin should be analyzed to quantify which elements are preserved and determine how much of the not-completely retained elements may be lost to metamorphic fluids. For simplicity, initial studies could focus on contact metamorphosed shales near intrusions, then the studies could be extended to variably metamorphosed basins.

It is not well understood how depositional setting affects the trace element content of pyrite. Most current studies focus on a few drill cores rather than a systematic sampling of cores from different sedimentary facies. The only current study that discusses trace element variation in pyrite from different depositional facies (Guy et al., 2010) suggests that relatively shallow facies will have lower trace element contents than deeper depositional facies. However, this study is on Mesoarchean rocks and it is not clear whether these results apply to post-Archean time periods when atmospheric oxygen levels are much higher and thus depositional flux are significantly different. Also, at these times, pyrite deposition might also be more related to volcanic processes than more recent times (i.e., post-GOE) when sulfate levels are much higher (Olson et al., 2019). To test how pyrite trace element content varies across a basin, a systematic set of samples can be taken from contemporaneous, sedimentary samples from different facies from a well-understood, non-metamorphosed basin. These data could then be used to assess the potential variability of the pyrite trace element content due to depositional facies. Initial steps have been made on this (Mukherjee et al., 2018b) with Cambrian shales in Australia but the results are yet to be published in full and the procedure should be repeated on multiple basins at different periods in Earth History.

Finally, little experimental work has been done to assess partition coefficients of different trace elements into pyrite at low temperatures. Swanner et al., 2019 investigated Co and Ni incorporation into pyrite and which is more representative of water chemistry from which it formed, but few other similar studies have been published. Thus, to fully understand the relationship between pyrite chemistry and its relationship to the chemistry of the water from which it formed, pyrite should be produced in solutions with different trace element compositions and the trace element content of the resultant pyrite should be analysed to obtain a set of partition coefficients.

While there is still more work required to refine the method, using pyrite trace element chemistry has many potential advantages. The ability to be preserved through metamorphism is perhaps the most important, especially in the Precambrian where researchers are frustrated by a vanishingly small number of basins that can be utilized to understand ocean and atmospheric chemistry at important periods of the evolution of our planet. The pyrite technique may provide glimpses into basins that currently are too high metamorphic grade for traditional proxies to be applied. The number of elements that can be accurately analyzed by LA-ICPMS and appear to reflect past ocean chemistry can give support to conclusions because they allow for the application of several lines of evidence to support interpretations. Finally, the ability of pyrite chemistry to identify the presence or lack of an overprint by hydrothermal fluids can help to resolve several disagreements whether a given redox sensitive trace element enrichment is due to changes in past ocean chemistry or later hydrothermal and/ or metamorphic overprint.

Key References

This volume provides an extensive overview of how pyrite forms in sedimentary settings.

Rickard, D. (2012) *Sulfidic Sediments and Sedimentary Rocks* (Elsevier) p. 801.

These papers are among the first to investigate sedimentary pyrite trace element abundance using methods other than LA-ICPMS.

Berner, Z. A., Puchelt, H., Nöltner, T. and Kramar, U. T. Z. (2013) Pyrite geochemistry in the Toarcian Posidonia Shale of south-west Germany: Evidence for contrasting trace-element patterns of diagenetic and syngenetic pyrites. *Sedimentology*, 60 548–573.

Huerta-Diaz, M. A., and Morse, J. W. (1990) A quantitative method for determination of trace metal concentrations in sedimentary pyrite. *Marine Chemistry* 29, 119–144.

Huerta-Diaz, M. A., and Morse, J. W. (1992) Pyritization of trace metals in anoxic marine sediments. *Geochimica et Cosmochimica Acta* 56, 2681–2702.

These papers were among the first to compile LA-ICPMS trace element data of pyrite and showed that it matches existing whole rock studies.

Gregory, D. D., Large, R. R., Halpin, J.A., Baturina, E. L., Lyons, T.W., Wu, S., Danyushevsky, L., Sack, P. J., Chappaz, A., and Maslennikov, V. V. (2015a) Trace Element Content of Sedimentary Pyrite in Black Shales. *Economic Geology* 110, 1389–1410.

Large, R. R., Halpin, J. A., Danyushevsky, L. V., Maslennikov, V. V., Bull, S. W., Long, J. A., Gregory, D. D., Lounejeva, E., Lyons, T. W., and Sack, P. J. (2014) Trace element content of sedimentary pyrite as a new proxy for deep-time ocean–atmosphere evolution. *Earth and Planetary Science Letters* 389, 209–220.

These papers provide examples where the proxy was used to identify oxygenation events at different times in Earth History.

Gregory, D. D., Large R. R., Halpin, J. A., Steadman, J. A., Hickman, A. H., Ireland, T. R., and Holden, P. (2015b) The chemical conditions of the late Archean Hamersley basin inferred from whole rock and pyrite geochemistry with Δ 33 S and δ 34 S isotope analyses. *Geochimica et Cosmochimica Acta* 149, 223–250.

Gregory, D. D., Lyons, T. W., Large, R. R., Jiang, G., Stepanov, A. S., Diamond, C. W., Figueroa, M. C., and Olin, P. (2017) Whole rock and discrete pyrite geochemistry as complementary tracers of ancient ocean carbonate succession from the Western Interior Seaway: Geochimica et Cosmochimica Acta, v. 186, pp. 277–95.

Hood, A. V., Planavsky, N. J., Wallace, M. W., and Wang, X. L., 2018, The effects of diagenesis on geochemical paleoredox proxies in sedimentary carbonates: Geochimica et Cosmochimica Acta, v. 232, pp. 265–87.

Janssen, D. J., Rickli, J., Quay, P. D. et al., 2020, Biological control of chromium redox and stable isotope composition in the surface ocean: Global Biogeochemical Cycles, v. 34, no. 1. https://doi.org/10.1029/2019GB006397

Johnson, C. A., and Xyla, A. G., 1991, The oxidation of chromium(III) to chromium(VI) on the surface of manganite (γ-MnOOH): Geochimica et Cosmochimica Acta, v. 55, no. 10, pp. 2861–6.

Johnson, T. M., and Bullen, T. D., 2004, Mass-dependent fractionation of selenium and chromium isotopes in low-temperature environments: Reviews in Mineralogy and Geochemistry, v. 55, no. 1, pp. 289–317.

Johnson, T. M., and DePaolo, D. J., 1994, Interpretation of isotopic data in groundwater-rock systems: Model development and application to Sr isotope data from Yucca Mountain: Water Resources Research, v. 30, pp. 1571–87.

Keeling, R. F., Kortzinger, A., and Gruber, N., 2010, Ocean deoxygenation in a warming world: Annual Review of Marine Science, v. 2, pp. 199–229.

Konhauser, K. O., Lalonde, S. V., Planavsky, N. J. et al., 2011, Aerobic bacterial pyrite oxidation and acid rock drainage during the Great Oxidation Event: Nature, v. 478, no. 7369, pp. 369–373.

Kump, L., 2008, The rise of atmospheric oxygen: Nature, v. 451, pp. 277–8.

Lyons, T. W., Reinhard, C. T., and Planavsky, N. J., 2014, The rise of oxygen in Earth's early ocean and atmosphere: Nature, v. 506, no. 7488, pp. 307–15.

Lyons, T. W., Werne, J. P., Hollander, D. J., and Murray, R. W., 2003, Contrasting sulfur geochemistry and Fe/Al and Mo/Al ratios across the last oxic-to-anoxic transition in the Cariaco Basin, Venezuela: Chemical Geology, v. 195, pp. 131–57.

McClain, C. N., and Maher, K., 2016, Chromium fluxes and speciation in ultra-mafic catchments and global rivers: Chemical Geology, v. 426, pp. 135–57.

Moos, S. B., and Boyle, E. A., 2019, Determination of accurate and precise chromium isotope ratios in seawater samples by MC-ICP-MS illustrated by analysis of SAFe Station in the North Pacific Ocean: Chemical Geology, v. 511, pp. 481–93.

Moos, S. B., Boyle, E. A., Altabet, M. A., and Bourbonnais, A., 2020, Investigating the cycling of chromium in the oxygen deficient waters of the Eastern Tropical North Pacific Ocean and the Santa Barbara Basin using stable isotopes: Marine Chemistry, v. 221. https://doi.org/10.1016/j.marchem.2020.103756

Nasemann, P. H., Janssen, D. J., Rickli, J. et al., 2020, Chromium reduction and associated stable isotope fractionation restricted to anoxic shelf waters in the Peruvian Oxygen Minimum Zone: Geochimica et cosmochimica acta, v. 285, pp. 207–24.

Oze, C., Bird, D. K., and Fendorf, S., 2007, Genesis of hexavalent chromium from natural sources in soil and groundwater: Proceedings of the National Academy of Sciences, v. 104, no. 16, pp. 6544–9.

Patterson, R. R., Fendorf, S., and Fendorf, M., 1997, Reduction of hexavalent chromium by amorphous iron sulfide: Environmental Science & Technology, v. 31, no. 7, pp. 2039–44.

Paulukat, C., Gilleaudeau, G. J., Chernyavskiy, P., and Frei, R., 2016, The Cr-isotope signature of surface seawater – A global perspective: Chemical Geology, v. 444, pp. 101–9.

Pereira, N. S., Vögelin, A. R., Paulukat, C. et al., 2015, Chromium isotope signatures in scleractinian corals from the Rocas Atoll, Tropical South Atlantic: Geobiology, v. 4, no. 1, p. 1–13.

Pereira, N. S., Voegelin, A. R., Paulukat, C. et al., 2016, Chromium-isotope signatures in scleractinian corals from the Rocas Atoll, Tropical South Atlantic: Geobiology, v. 14, no. 1, pp. 54–67.

Planavsky, N. J., Cole, D.B., Isson, T.T. et al., 2018, A case for low atmospheric oxygen levels during Earth's middle history: Emerging Topics in Life Sciences, p. ETLS20170161.

Planavsky, N. J., Reinhard, C. T., Wang, X. L. et al., 2014, Low Mid-Proterozoic atmospheric oxygen levels and the delayed rise of animals: Science, v. 346, no. 6209, pp. 635–8.

Reinhard, C. T., Planavsky, N. J., Robbins, L. J. et al., 2013, Proterozoic ocean redox and biogeochemical stasis: Proceedings of the National Academy of Sciences USA, v. 110, pp. 5357–62.

Reinhard, C. T., Planavsky, N. J., Wang, X. et al., 2014, The isotopic composition of authigenic chromium in anoxic marine sediments: A case study from the Cariaco Basin: Earth and Planetary Science Letters, v. 407, pp. 9–18.

Remmelzwaal, S. R. C., Sadekov, A. Y., Parkinson, I. J. et al., 2019, Post-depositional overprinting of chromium in foraminifera: Earth and Planetary Science Letters, v. 515, pp. 100–11.

Richard, F. C., and Bourg, A. C. M., 1991, Aqueous geochemistry of chromium: A review: Water Research, v. 25, no. 7, pp. 807–16.

Rickli, J., Janssen, D. J., Hassler, C., Ellwood, M. J., and Jaccard, S. L., 2019, Chromium biogeochemistry and stable isotope distribution in the Southern Ocean: Geochimica Et Cosmochimica Acta, v. 262, pp. 188–206.

Rodler, A. S., Frei, R., Gaucher, C., and Germs, G. J. B., 2016, Chromium isotope, REE and redox-sensitive trace element chemostratigraphy across the late Neoproterozoic Ghaub glaciation, Otavi Group, Namibia: Precambrian Research, v. 286, pp. 234–49.

Rudnick, R. L., and Gao, S., 2003, Composition of the continental crust: Treatise of Geochemistry, v. 3, pp. 1–64.

Saad, E. M., Wang, X. L., Planavsky, N. J., Reinhard, C. T., and Tang, Y. Z., 2017, Redox-independent chromium isotope fractionation induced by ligand-promoted dissolution: Nature Communications, v. 8, pp. 1–10.

Sander, S., and Koschinsky, A., 2000, Onboard-ship redox speciation of chromium in diffuse hydrothermal fluids from the North Fiji Basin: Marine Chemistry, v. 71, no. 1–2, pp. 83–102.

Sander, S., Koschinsky, A., and Halbach, P., 2003, Redox speciation of chromium in the oceanic water column of the Lesser Antilles and offshore Otago Peninsula, New Zealand: Marine and Freshwater Research, v. 54, no. 6, pp. 745–54.

Schauble, E., Rossman, G. R., and Taylor Jr, H. P., 2004, Theoretical estimates of equilibrium chromium isotope fractionations: Chemical Geology, v. 205, no. 1–2, pp. 99–114.

Scheiderich, K., Amini, M., Holmden, C., and Francois, R., 2015, Global variability of chromium isotopes in seawater demonstrated by Pacific, Atlantic, and Arctic Ocean samples: Earth and Planetary Science Letters, v. 423, pp. 87–97.

Schoenberg, R., Zink, S., Staubwasser, M., and von Blanckenburg, F., 2008, The stable Cr isotope inventory of solid Earth reservoirs determined by double spike MC-ICP-MS: Chemical Geology, v. 249, no. 3–4, pp. 294–306.

Scholz, F., Severmann, S., McManus, J. et al., 2014, On the isotope composition of reactive iron in marine sediments: Redox shuttle versus early diagenesis: Chemical Geology, v. 389, pp. 48–59.

Shaw, T. J., Gieskes, J. M., and Jahnke, R. A., 1990, Early diagenesis in differing depositional environments: The response of transition metals in pore water: Geochimica et Cosmochimica Acta, v. 54, pp. 1233–46.

Sial, A. N., Campos, M. S., Gaucher, C. et al., 2015, Algoma-type Neoproterozoic BIFs and related marbles in the Serido Belt (NE Brazil): REE, C, O, Cr and Sr isotope evidence: Journal of South American Earth Sciences, v. 61, pp. 33–52.

Sperling, E. A., Halverson, G. P., Knoll, A. H., Macdonald, F. A., and Johnston, D. T., 2013, A basin redox transect at the dawn of animal life: Earth and Planetary Science Letters, v. 371, pp. 143–55.

Sun, Z. Y., Wang, X. L., and Planavsky, N., 2019, Cr isotope systematics in the Connecticut River estuary: Chemical Geology, v. 506, pp. 29–39.

Toma, J., Holmden, C., Shakotko, P., Pan, Y., and Ootes, L., 2019, Cr isotopic insights into ca. 1.9 Ga oxidative weathering of the continents using the Beaverlodge Lake paleosol, Northwest Territories, Canada: Geobiology, v. 17, no. 5, pp. 467–89.

Towe, K. M., 1970, Oxygen-collagen priority and early Metazoan fossil record: Proceedings of the National Academy of Sciences of the United States of America, v. 65, no. 4, pp. 781–788.

Wang, X. L., Glass, J. B., Reinhard, C. T., and Planavsky, N. J., 2019, Species-dependent chromium isotope fractionation across the Eastern Tropical North Pacific Oxygen Minimum Zone: Geochemistry Geophysics Geosystems, v. 20, no. 5, pp. 2499–514.

Wang, X. L., Planavsky, N. J., Hull, P. M. et al., 2017, Chromium isotopic composition of core-top planktonic foraminifera: Geobiology, v. 15, no. 1, pp. 51–64.

Wang, X. L., Planavsky, N. J., Reinhard, C. T. et al., 2016a, Chromium isotope fractionation during subduction-related metamorphism, black shale weathering, and hydrothermal alteration: Chemical Geology, v. 423, pp. 19–33.

Wang, X. L., Reinhard, C. T., Planavsky, N. J. et al., 2016b, Sedimentary chromium isotopic compositions across the Cretaceous OAE2 at Demerara Rise Site 1258: Chemical Geology, v. 429, pp. 85–92.

Wei, W., Frei, R., Chen, T. Y. et al., 2018, Marine ferromanganese oxide: A potentially important sink of light chromium isotopes?: Chemical Geology, v. 495, pp. 90–103.

Wu, W. H., Wang, X. L., Reinhard, C. T., and Planavsky, N. J., 2017, Chromium isotope systematics in the Connecticut River: Chemical Geology, v. 456, pp. 98–111.

Zink, S., Schoenberg, R., and Staubwasser, M., 2010, Isotopic fractionation and reaction kinetics between Cr(III) and Cr(VI) in aqueous media: Geochimica Et Cosmochimica Acta, v. 74, no. 20, pp. 5729–45.

Cambridge Elements ≡

Elements in Geochemical Tracers in Earth System Science

Timothy Lyons
University of California

Timothy Lyons is a Distinguished Professor of Biogeochemistry in the Department of Earth Sciences at the University of California, Riverside. He is an expert in the use of geochemical tracers for applications in astrobiology, geobiology and Earth history. Professor Lyons leads the 'Alternative Earths' team of the NASA Astrobiology Institute and the Alternative Earths Astrobiology Center at UC Riverside.

Alexandra Turchyn
University of Cambridge

Alexandra Turchyn is a University Reader in Biogeochemistry in the Department of Earth Sciences at the University of Cambridge. Her primary research interests are in isotope geochemistry and the application of geochemistry to interrogate modern and past environments.

Chris Reinhard
Georgia Institute of Technology

Chris Reinhard is an Assistant Professor in the Department of Earth and Atmospheric Sciences at the Georgia Institute of Technology. His research focuses on biogeochemistry and paleoclimatology, and he is an Institutional PI on the 'Alternative Earths' team of the NASA Astrobiology Institute.

About the Series

This innovative series provides authoritative, concise overviews of the many novel isotope and elemental systems that can be used as "proxies" or "geochemical tracers" to reconstruct past environments over thousands to millions to billions of years – from the evolving chemistry of the atmosphere and oceans to their cause-and-effect relationships with life.

Covering a wide variety of geochemical tracers, the series reviews each method in terms of the geochemical underpinnings, the promises and pitfalls, and the "state-of-the-art" and future prospects, providing a dynamic reference resource for graduate students, researchers and scientists in geochemistry, astrobiology, paleontology, paleoceanography, and paleoclimatology.

The short, timely, broadly accessible papers provide much-needed primers for a wide audience – highlighting the cutting-edge of both new and established proxies as applied to diverse questions about Earth system evolution over wide-ranging time scales.

Cambridge Elements ≡

Elements in Geochemical Tracers in Earth System Science

Elements in the Series